True Identity of Dark Matter!?

"Proto-Particle" Produce Elementary Particles

Hideo Asawa

Contents

Preface

The content of this book is a philosophical approach to the unified theory of the universe. In Newton's era, physics was called natural philosophy. So, it is possible that philosophy can lead science. At least I think this book has some hints.

In my view, philosophy established by Dr. Kenzo Yamamoto - namely a 6-dimensional principle - represents the truth of the universe. I will briefly explain the 6-dimensional principle in this book. To prove that it is really the truth of the universe, I have thought that it would be good if the fundamentals of cosmophysics could be explained in the 6-dimensional principle.

Also, I took inspiration from Plato who said "The reality people seeing is only the shadow of the *Idea* world (the true reality world)." Because elementary particles are invisible, we can only look at trajectories in a mist box or observe them with a sensor response. In other words, we seem to observe a shadow of the natural world.

From there, I came to think that an elementary particle considered to be a basic material of the universe may be a phenomenon as a shadow of a root body in true existence.

In other words, if the 6-dimensional principle is the truth of the universe, the true root body is consistent with the 6-dimensional principle, and it should be possible to explain the appearance of elementary particles by the 6-dimensional principle.

To conclude first, if the universe is six-dimensional, the root of the universe can be explained by the "Proto-particle" (philosophical tetrahedron) I named. And the elementary particles that make up substances can be explained in a unified way as phenomena of "Proto-particles".

By the way, it was end of 2014 when I came up with the "Proto-particle" (philosophical tetrahedron). I published it on my homepage in the spring of 2015, but it was difficult to understand, so I revised it in various ways and it became this book.

Also, if the 6-dimensional theory is the truth of the universe, the 6-dimensional theory should be applicable to the life and mind born in the universe. For that, please read the following my book.
Mind Is Neutrino!? It Gives Evolution and Psychic Power

1. Elementary particles that make up substances can be unified by "Proto-particle"

1-1. What are elementary particles

Elementary particles are the most basic particles considered to constitute everything in the universe. Several kinds of elementary particles gather and exert force on each other, creating protons, neutrons, etc. They combine to form various atoms, and atoms gather to form various materials. Furthermore, matter gathers to create stars and planets, to create galaxies, and to create the vast universe.

Please look at the figure below, titled "Standard Model of Elementary Particles." (Quoted from Wikipedia "Standard Model." I inserted the black vertical line there.) The figure shows all currently known elementary particles and their classification.

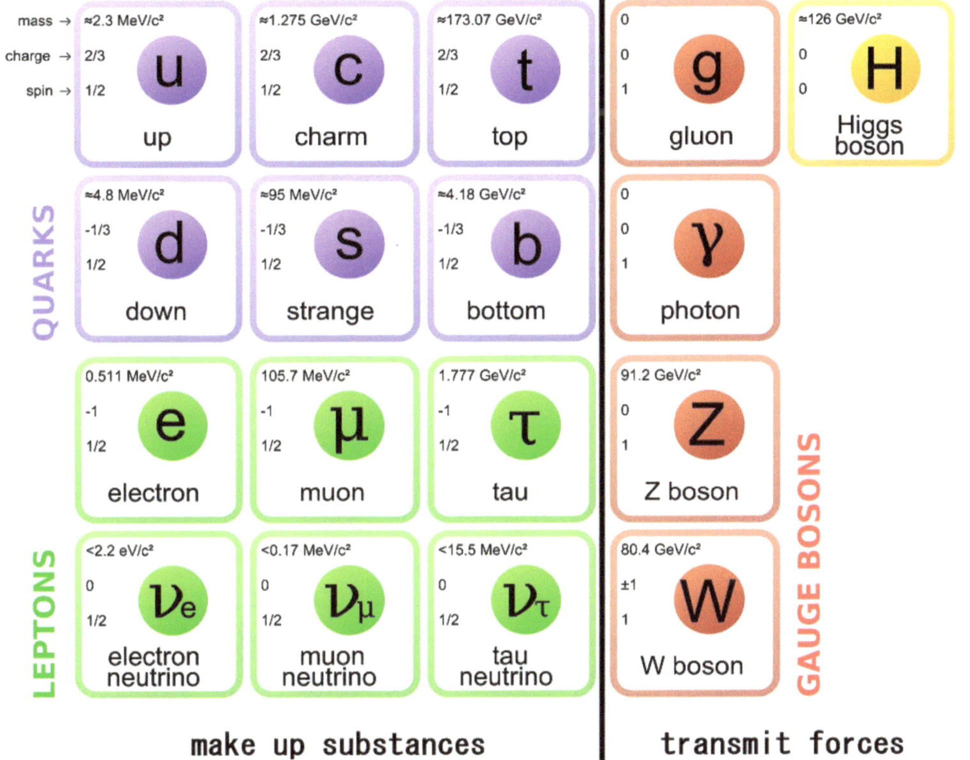

The elementary particles that make up substances are quarks (blue) and leptons (green), shown in the 3 columns on the left side of the vertical black line. There are 6 types of quarks and leptons each and they are classified into 4 horizontal groups.

In addition, as shown on the right side of the vertical line in the figure, there are 4 types of elementary particles called bosons (red) that transmit forces.

According to the Big Bang theory, the beginning of the universe was like one single point. In other words, these elementary particles which seem to exist separately were originally one thing, so there is a possibility that these particles can be described in a unified way.

In physics, it may be difficult to solve the unified theory of the universe, as it is particular about mathematical formulas. However, I reckon that philosophy may be able to provide a conceptual explanation.

1-2. Think about elementary particles of matter

I will discuss only the elementary particles that make up the basic substance. Elementary particles that transmit forces are excluded here because they are force or energy that are generated concomitantly with the elementary particles that make up a substance and are not considered to be basic substances.

That is, I will only consider the elementary particles of the following figure.

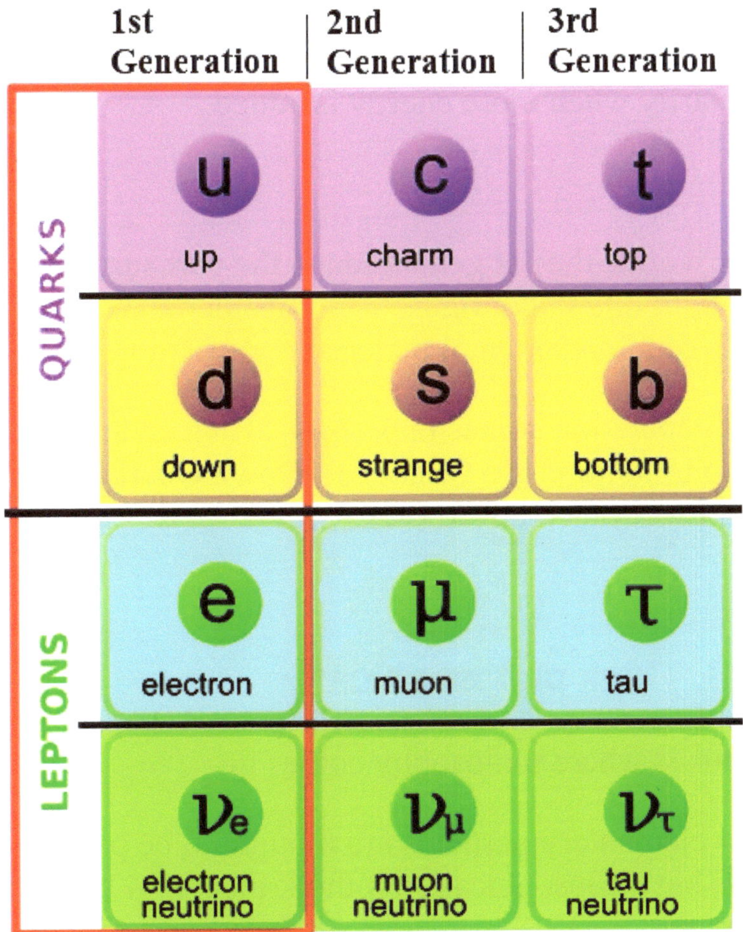

The diagrams are color coded for each type. As mentioned earlier, in the vertical direction of the figure, there are 2 types which are roughly classified, quark and lepton, each is further classified into 2, so there are 4 types in each column.

For each type, in the horizontal direction of the figure, the 1st column is called the 1st generation, the 2nd column is the 2nd generation, and the 3rd column is the 3rd generation.

The 2nd generation appears for a short time when high energy is added in experiments. The 3rd generation appears momentarily when more energy is added.

In other words, the 1st generation -- the part surrounded by the red bold line in the figure -- is considered elementary particles of a fundamental material composition.

The 1st generation comes in 4 types, arranged in the figure from top to bottom: Up Quark (u), Down Quark (d), Electron (e), and Electron Neutrino (*ν*e).

1-3. Think philosophically

Well, this is where philosophy comes in.

These 4 elementary particles are thought to be particles in physics, but philosophically let think them as 4 kinds of phenomena.

In the first place, because elementary particles are identified by observing various phenomena, elementary particles may be regarded as phenomena.

Before the Big Bang, the universe was like a single point. This is a collection of the root bodies of the universe, and for convenience, I call the root body "Proto-particle". I think that 4 kinds of elementary particles were generated as 4 kinds of phenomena from this "Proto- particle".

Did the "Proto- particle" split into 4? No, I think that the "Proto- particle" is 1 thing and cannot be divided. So how does 1 thing show 4 kinds of phenomena?

It is possible if we think that 1 thing has 4 sides. I think that we observe the 4 sides of a "Proto-particle" and recognize that there are 4 types of elementary particles.

There is a famous Indian fable. The blind people touch different parts of the elephant, and each person says that the elephant is a wall, a pillar or a rope, etc. I think the situation is similar to that fable. Namely, we observe some side of a "Proto-particle" and recognize them as quarks, electrons, or so on.

Well, what is 1 object with 4 aspects? It is a tetrahedron -- a triangular pyramid. However, it's only expressed in a way we can recognize, and we don't know how it really is. Therefore, I will call it "philosophical tetrahedron", but it is "Proto-particle".

1-4. Show elementary particles in the "Proto-particle"

I will dig into the "Proto-particle". I will say it persistently, but keep in mind that this tetrahedron is not a material structure, but a shape that can be recognized by humans.

A tetrahedron of the "Proto-particle" is illustrated as follows.

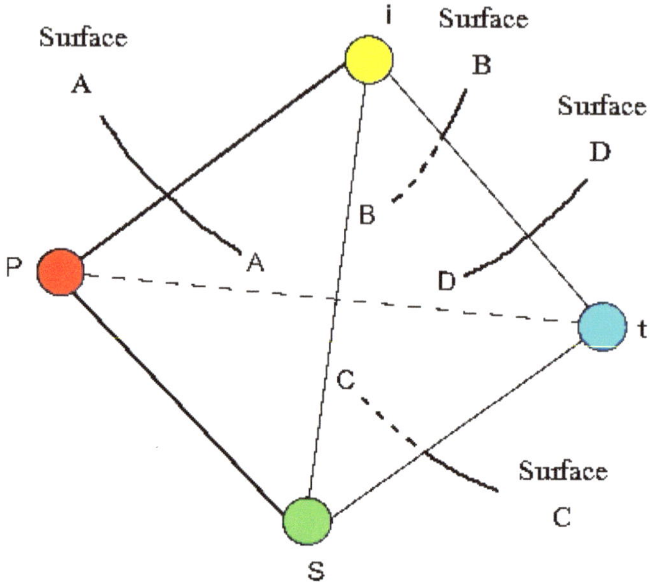

I named each of the 4 side surfaces A, B, C and D.

Also, I set 4 vertices as p, s, t, and i, but I will postpone the explanation of these 4 vertices. And, I colored the 4 vertices for identification.

These 4 planes are phenomena surfaces.

For example, the figure below shows the state in which surface A has caused the phenomenon, and we can see only surface A in color.

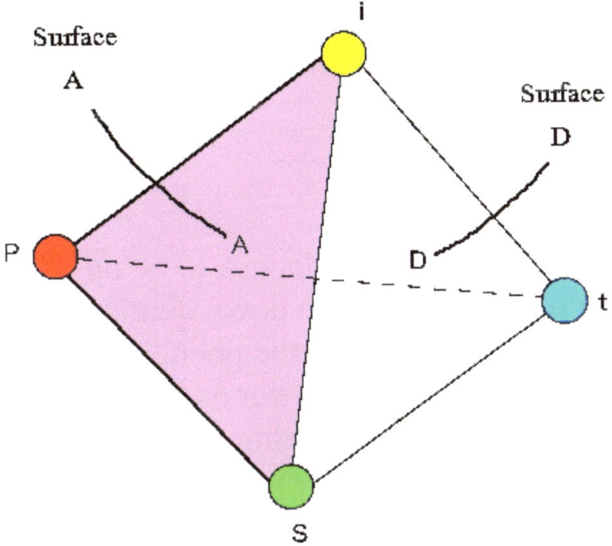

Also, when the surface D causes a phenomenon, it looks like the figure below, and only the surface D appears to us in color.

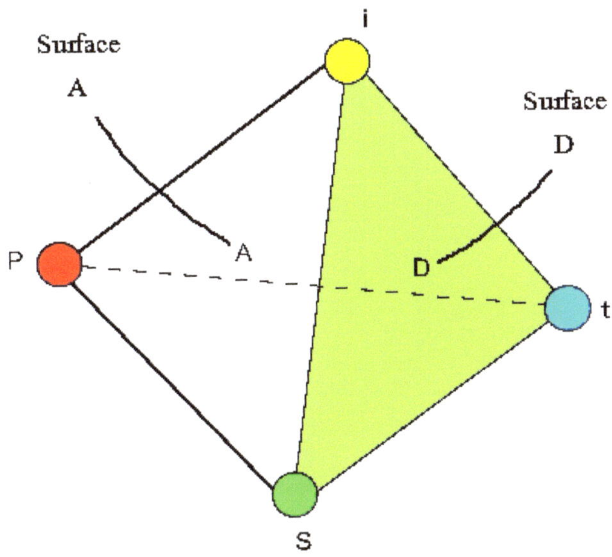

The fact that a surface appears colored means that the surface can be recognized as a phenomenon, that is, it can be observed.

If I set the phenomenon of surface A to be up quark, the phenomenon of surface B to be down quark, the phenomenon of surface C to be electron and the phenomenon of surface D to be electron neutrino, they appear as in the table below. The colors themselves have no meaning. Please refer to Chapter 2 about why I applied elementary particles that way.

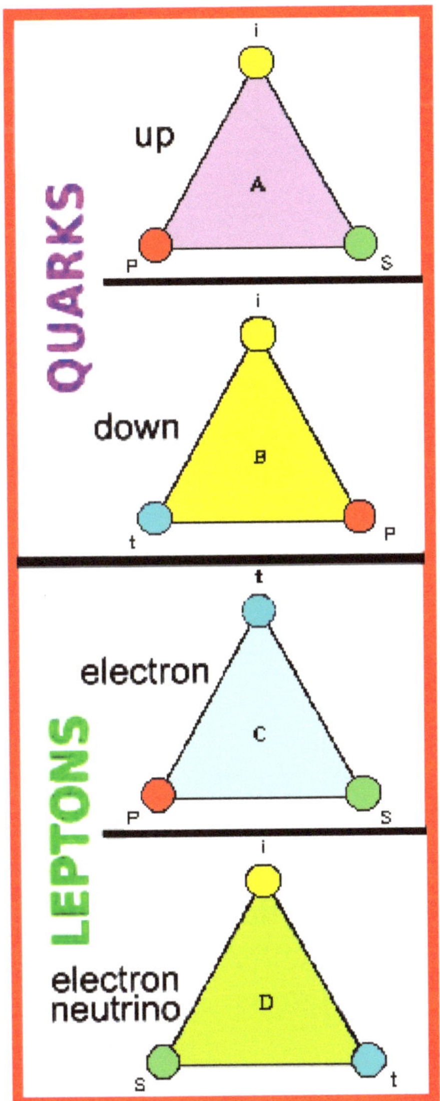

The following figure shows this result in the "Proto-particle".

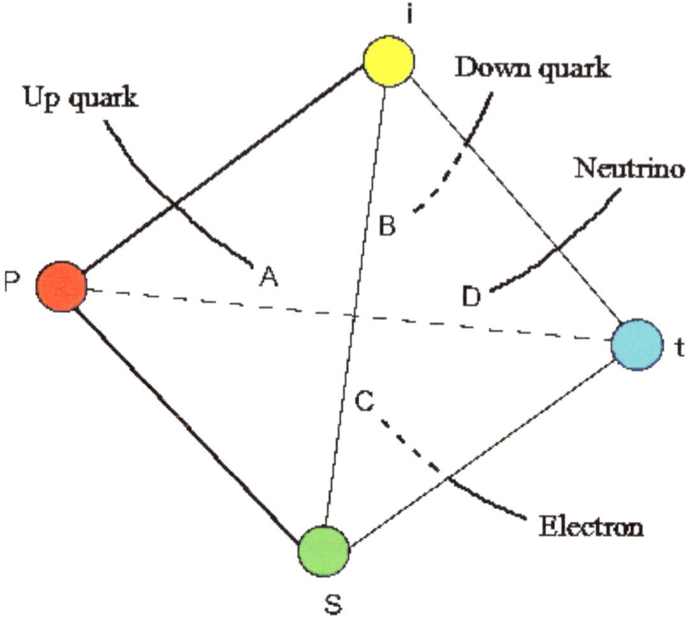

1-5. Think about the "Proto-particle"

It is said that quark has 3 properties called "Color". Looking at the face A and face B of the quark in the "Proto-particle" shown above, each has 3 colored vertices, which has possibility to correspond to the 3 "Colors" of the quark.

If so, it is possible that there are also 3 "Colors" in electron and electron neutrino which are leptons, but it is not recognized at present.

By the way, if we distinguish tetrahedrons by placing symbols on the faces and vertices, there are isomers that look like mirror images of the original tetrahedrons. It shows as follows.

I made a color on surface A' so that it is easier to compare with the figure shown above, where the phenomenon appears on surface A.

Isomer

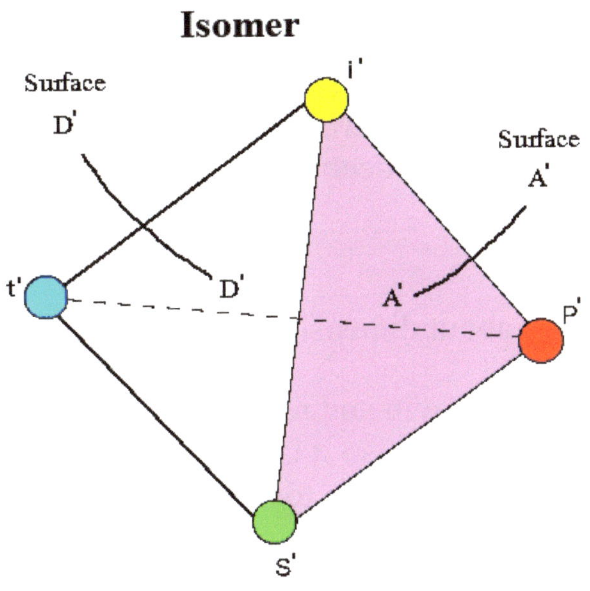

Origin

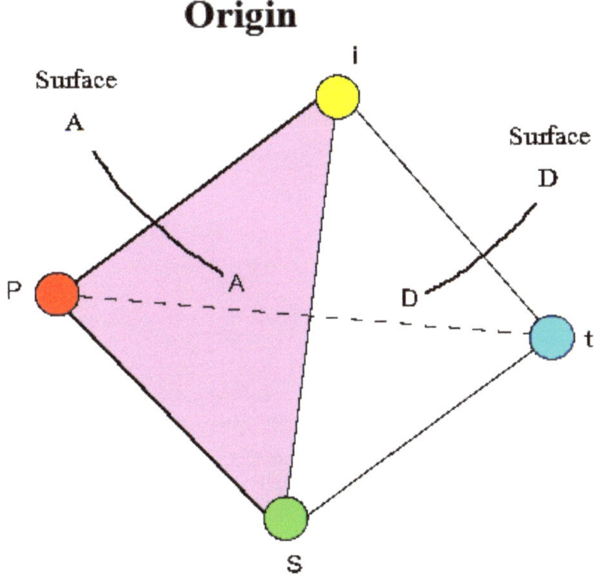

Even in this isomer of the "Proto-particle", there should be elementary particles as phenomena of faces A', B', C' and D'.

It is unknown whether they are antiparticles or supersymmetric particles for up quarks, down quarks, electrons and electron neutrinos, but there is a possibility that one of them may be applicable.

Well, let's think about dimensions here.

The space that we live in and recognize has 3 dimensions -- length, width and height, so it is a 3-dimensional space. By the way, Einstein added time to it and said that the universe is 4-dimensional.

Speaking of 3 dimensions, you should have seen the following diagram in mathematics.

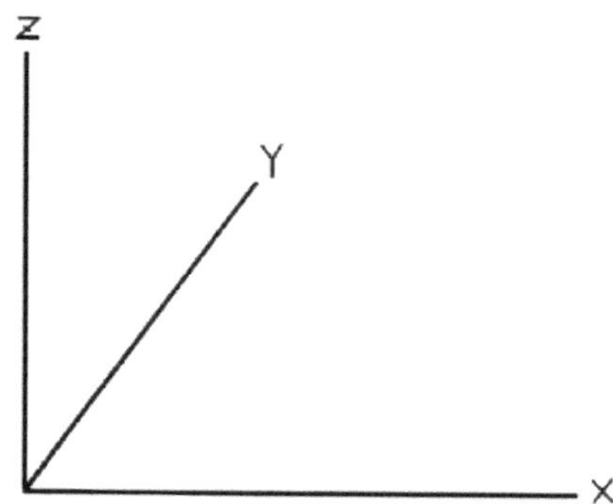

If there is only 1 X-axis, it is 1-dimensional, and if there are 2

X-axis and Y-axis, it is 2-dimensional, but the figure shows 3-dimensional because there are 3 axes, X, Y, and Z.

Please think that an axis represents a direction or arm that can extend and contract freely.

Well, what is the number of dimensions in a tetrahedron of "Proto-particle"? How many arms (lines) are there?

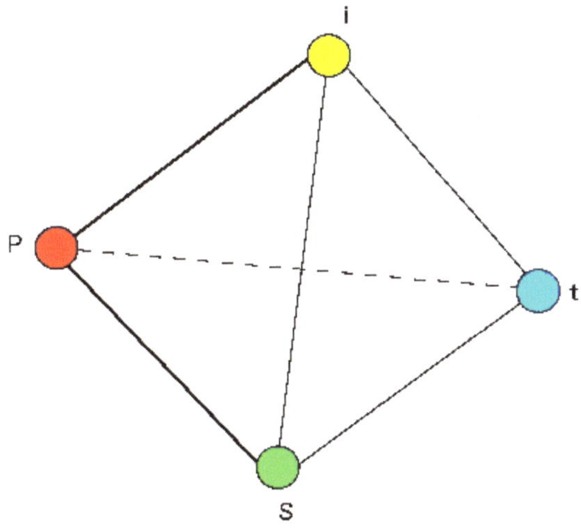

Yes, there are 6 arms (axes). The "Proto-particle" is 6-dimensional.

I will think about isomer again here.

Let us assume that the "Proto-particle" and its isomer are joined as shown in the following figure.

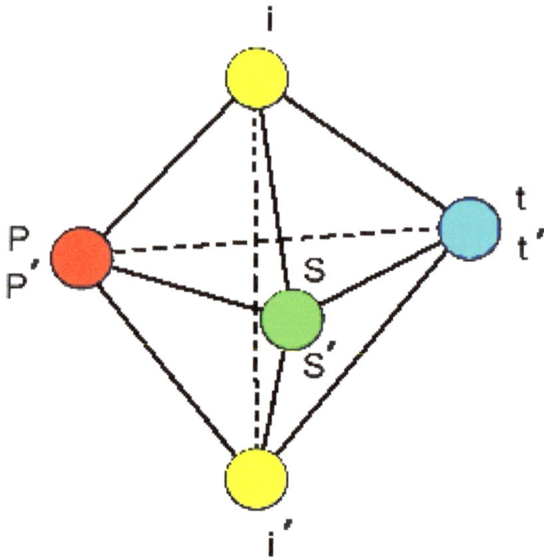

Well, how many dimensions this complex has? Yes, it has 10 dimensions because there are 10 arms. There are two 6-dimensional tetrahedra, but they join to make 1 complex with 10 dimensions.

By the way, the superstring theory of physics postulates a possibility to explain the universe in a unified way. According to the superstring theory, the universe has 10 or 11 dimensions, with 6 dimensions rounded in. This binding model fits well.

1-6. The universe is 6-dimensional

I do not know whether the bound state of the "Proto-particle" and its isomer is normal or special. However, since the

"Proto-particle", which is the root body of the universe, has 6 dimensions, the fundamental of the universe should be 6-dimensional.

As mentioned above, if the root body of the universe is the "Proto-particle" (philosophical tetrahedron), elementary particles that make up substances of the universe can be unified as phenomena of "Proto-particles". In short, the philosophical conclusion is that the universe is 6-dimensional.

2. Philosophy of "Proto-particle" - intention produces gravity

2-1. Philosophy of the 6-dimensional principle

According to the 6-dimensional principle, the philosophical theory insisted by Dr. Kenzo Yamamoto (November 12, 1912 - August 7, 2007), there are 4 elements in human perception and world events and phenomena: space, time, power (energy) and intention (direction). Of course, they cannot be separated because they are originally one.

Space has 3 dimensions: height, width and depth, while time, power and intention each has 1 dimension. So, there are 6 dimensions in the 4 elements. Hence it is called 6-dimensional principle.

This can be visualized as shown in the figure below.

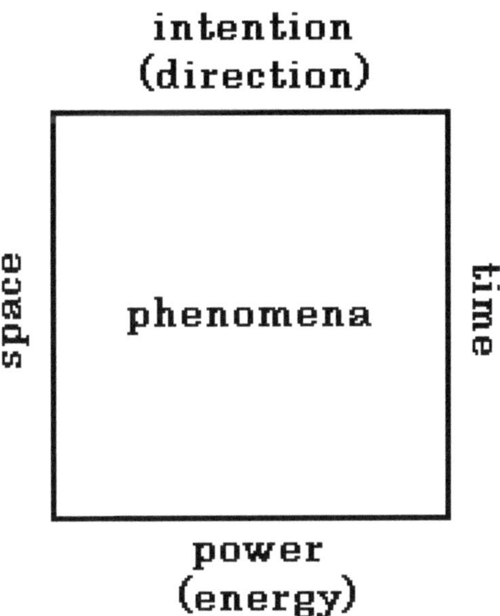

Ideally, people shouldn't be biased towards any of these 4 elements, but unfortunately most people's ideas are biased towards any.

In the world, a culture of thinking with easy-to-understand antinomy such as winning and losing, good and evil, pros and cons, yin and yang, high and low, 1 or 0, etc. has developed, but there are actually 4.

As a simple example, we often divide the world into west and east, but we need 4 in reality, including north and south.

Furthermore, when applying this 6-dimensional principle to human thought, for example, if you are biased towards space or power, you will be hedonism (such as good if happy now) or materialism (such as money or power is all).

Conversely, if you are biased towards time or intention, you will be fundamentalism (such as absolute habits) or spiritualism (such as superstitious religion).

And, when biased like these, people make mistakes in their lives by fighting for money or religion, or ignoring medicine and being too late.

So, the center, having no biases toward either, is our ideal position.

Changing the representation in the above figure, it will be as shown in the following figure.

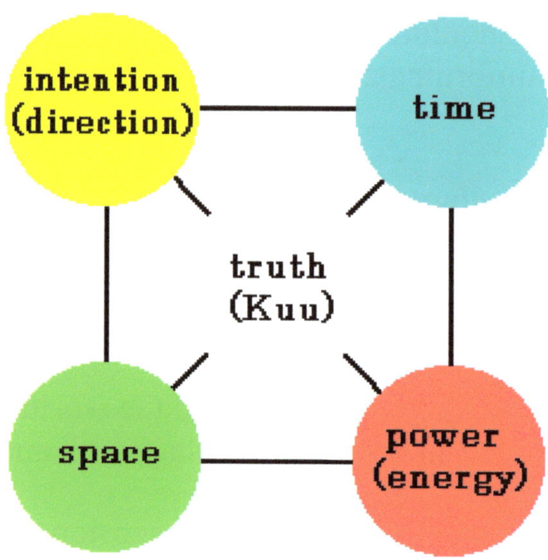

The truth is in the center, not biased toward any element. The middle "Kuu" will be explained in Chapter 3.

2-2. Why is it 6-dimensional rather than 4-dimensional?

Einstein concluded that the universe has 4 dimensions, consisting of space (3-dimensional) and time. Why is it need 2 more dimensions, power and intention?

Dr. Yamamoto discovered it while working in a rice field, so I will explain it in that situation.

If in the 4-dimensional world, you want to work with a hoe in a rice field, but you keep standing like a scarecrow in that space and only time passes, and you cannot do anything. In order for you to work with a hoe, you need the power to move the tool and the will to move it. Dr. Yamamoto considers this power and will (intention) as 2 different dimensions.

He concluded that the universe is 6 dimensions, which is 4 dimensions (space and time) plus that 2 dimensions (power and intention), and that it is the minimum number of dimensions required.

2-3. Power should be the 5th dimension

It is strange that power is not considered as a dimension in physics.

Regarding the motion of an object, the physics formula is:

<Force> = <Mass> x <Acceleration>

Acceleration is a value determined by the 4 dimensions of position (space) and time.

Then, how about mass?

After all, if we do not know acceleration or gravity and force or energy, we cannot determine mass.

Dimension is the minimum required measure to represent an event. Namely, I think it is necessary to regard force or energy as the 5th dimension.

On the other hand, in physics, there is a theory that the universe is 5-dimensional. And, as an explanation of the 5th dimension, it is said that there is a parallel universe, but isn't it more realistic to think of power as the 5th dimension, rather than fantasizing about a parallel universe?

2-4. Intention (direction) is also a dimension of the universe

As Dr. Yamamoto thought, intention (direction) is important in expressing the events of the world.

No event will occur unless you determine what and how, or in which direction. In mathematics, we think of a vector (direction with size), but it is not treated as a dimension. However, intention (direction) is also the minimum necessary physical scale to represent an event.

In other words, I think we need to think about the intention or direction as the 6th new dimension.

According to research by Dr. Yamamoto, in the 6-dimensional world, a material phenomenon without being limited by space and time is happened by the effect of intention, and that phenomenon can exceed the speed of light. Then, it will be possible that mysterious quantum entanglement or quantum teleportation which instantly transmits information between elementary particles far apart at the universe level.

I define to this ability of intention as "long-distance instantaneous communication capability between elements of intention," as described after.

I think that we can contribute to the development of physics by studying the action on matter with the intention as the 6th dimension.

2-5. Changing the expression becomes a philosophical tetrahedron

The figure below is a planar representation of the 6-dimensional principle described in 2-1.

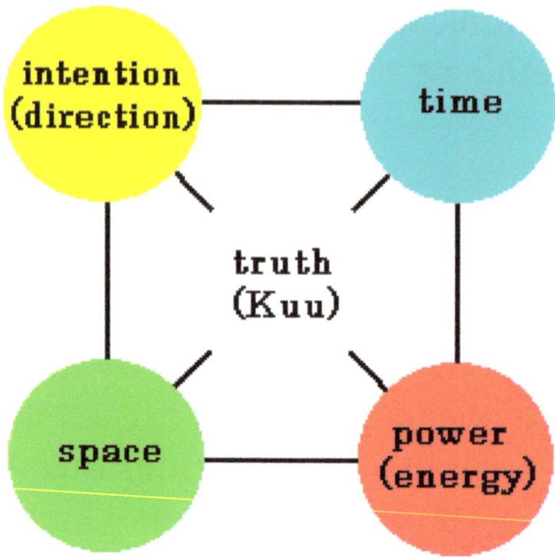

Changing it to a 3-dimensional expression will result in a tetrahedron (triangular pyramid) as shown below.

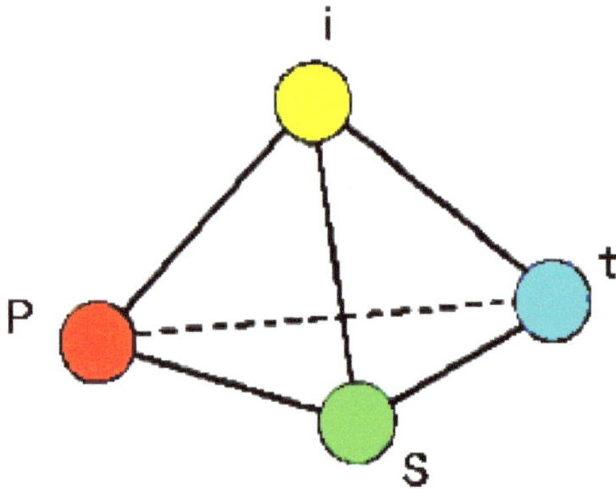

Here I express space as s, time as t, power as p, and intention as i. 4 colors are used for identification, they have no meaning in themselves.

Since this is a structure considered in terms of philosophical theory, I will call it a (6-dimensional) philosophical tetrahedron.

In the following there is a part that overlaps with Chapter 1, but it is necessary for philosophical explanation.

2-6. Phenomenon faces of a philosophical tetrahedron

According to the 6-dimensional principle, since any event / phenomenon in the world (universe) is a complex of 4 elements, in the following figure of a philosophical tetrahedron, A, B, C and D is each considered as a phenomenon surface in which elements are biased (not all 4).

So, the elementary particles that make up the matter in the universe be able to consider to be phenomena on each side face.

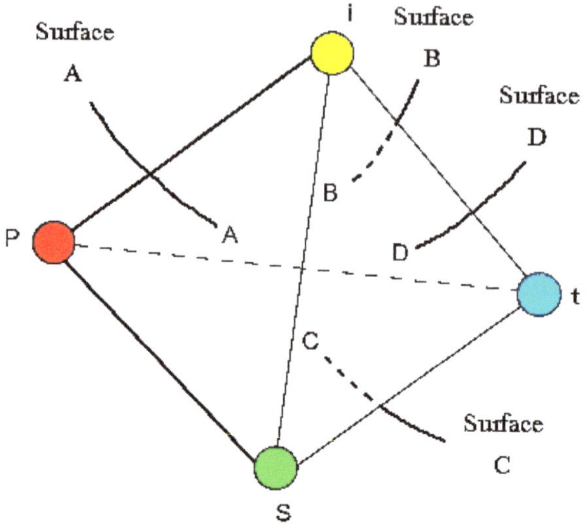

The relationship between each face of a philosophical tetrahedron and elementary particles is described in the following figure (same as presented in Chapter 1).

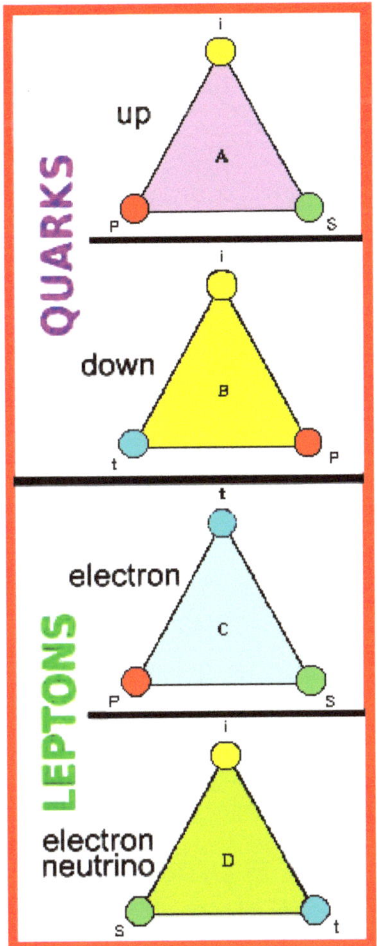

If the figure is changed into a 3-dimensional expression, it will look like this:

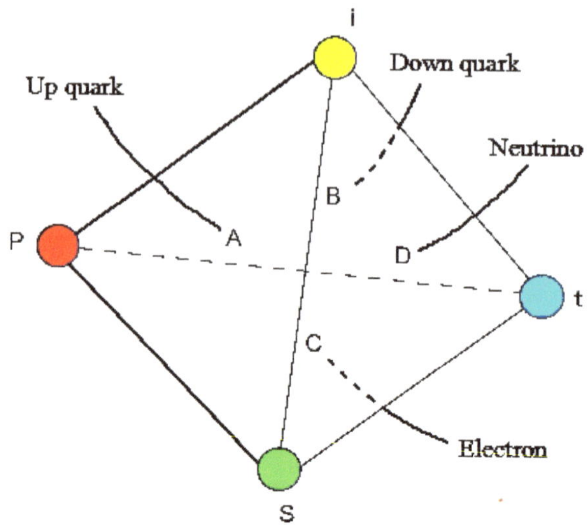

This philosophical tetrahedron is the "Proto-particle" in Chapter 1.

2-7. Elements of a philosophical tetrahedron, and elementary particles

In the figure above, the relationship between each phenomenon surface and the elementary particles is not randomly arranged by me. It takes into consideration the characteristics of each element (the vertex of the figure).

Regarding space, time, force, and intention of the 4 elements, I considered the dominant characteristics at the elementary particle level as follows:

* Space (s): Large, Wide

* Time (t): Long life

* Power (p): Strong force, Electromagnetic force, Large mass

* Intention (i): Instantaneous long-distance communication capability, Gravity, Memory

I will discuss gravity of the characteristic of the intention in the next section.

Considering these characteristics, in the case of quarks, since strong force acts, it contains p (power), and since it is affected by gravity, it also contains i (intention), so I place quarks as surface A and surface B.

Regarding electron, since strong force such as electromagnetic force acts, I place it as surface C, which contains p (power).

Since neutrino has a small mass and no electromagnetic force, I place it as surface D, which does not contain p (power).

By the way, what happens to the element that are not included in each phenomenon plane?

As mentioned earlier, the 4 elements -- space, time, power and intention -- cannot be separated. Therefore, since it is always a philosophical tetrahedron, element not included in the phenomenon plane is also connected and should exert some influence.

2-8. Intention elements produce gravity

Dr. Yamamoto considered that there is "intention quarks" apart from material quarks and that "intention quarks" have memory ability. "Intention quarks" differs from material quarks in that they have an element of intention.

This "intention quark" assumption is inconsistent in that the inseparable 6-dimensional elements are separated, so I disagree, but agree that the intention element has memory.

If the intention element (i) (refer figure below) in the "Proto-particle" as the philosophical tetrahedron also has a memory, it should have the memory like a sense of unity that it was in a state of a single point just before the Big Bang at the beginning of the universe.

And due to that memory, all the "Proto-particles" that make up the present universe call each other by their communication capability such as quantum entanglement, and try to become one again.

I think this action produces the mutually attractive force, that is, gravity.

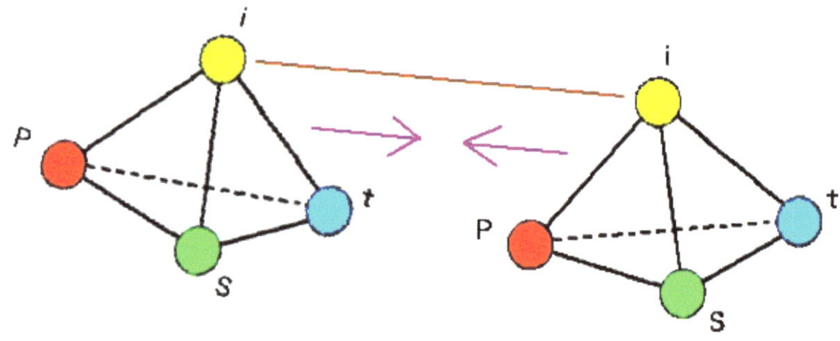

In physics, it is thought that forces act on each other with a catch ball of elementary particles. Gravity was also thought to be born of a catch ball of particles called graviton. However, in the catch ball we do, when we receive the ball, we are pushed by its kinetic energy, but when receiving the flying particle, it is difficult to understand that it pulls matter in reverse.

In order for gravitons to generate gravity, it is necessary to generate a large amount of them and fly in all directions at all times, although the mechanism is unknown, but I think it is more appropriate to think that the space is distorted and gravity is generated.

Einstein argued that the distortion of space-time causes gravity, which does not contradict the 6-dimensional principle. Because 4 elements as space, time, power, intention are inseparable and affect each other. Namely, the intention elements do not attract each other alone, but affects other elements such as distorting space-time and generating force. It occurs in the "Proto-particles" that fill the universe.

I thus hypothesized that communication capabilities such as quantum entanglement of the intention elements would create gravity. Since the phenomenon of quantum entanglement is a change in characteristics, it propagates instantaneously only by the communication ability of the intention element, but since gravity is a force, the power element also affects it.

Therefore, I think that the propagation speed of gravity is limited to the speed of light in the material world.

Dr. Kenzo Yamamoto was thinking before me that the element of intention produces gravity. He wrote in his book titled "6-Dimensional Dialectic Method":

"String-like shape before materialization [same as those constituting consciousness] ... I think that the intention of energy that the string is toward the center is gravity. ... The whole, part and gravity are simultaneous."

Prior to the introduction of the “intention quark” model mentioned earlier, Dr. Yamamoto considered something like tiny strings with the string theory or the superstring theory as a hint.

In the first place, the tiny strings in the string theory had been proposed as a concept to explain for the contradictions in cosmological formulas. It is similar to the philosophical tetrahedron in that it is a concept.

2-9. "Proto-particle" (philosophical tetrahedron) resemble "Proto-consciousness"

I learned later, about 20 years ago, there was a hypothesis that thought of what was named "Proto-consciousness" as "something like of component of the universe."

In summary, Dr. Stuart Hameroff, MD and director of the Center for Consciousness Studies at the University of Arizona, thought:

There are microtubules in the brain, and they work like a quantum computer. And it is making consciousness (mind) called "Quantum consciousness".

By quantum entanglement, "Quantum consciousness" is connected to "things like a component of the universe", that is, "Proto-consciousness" spreading in the universe.

By that, after the death of a person, that "Quantum consciousness" spreads into the universe.

Moreover, "Proto-consciousness" seems to exist from the Big Bang.

This "Proto-consciousness" is similar to the following points of the "Proto-particle" (philosophical tetrahedron). Below, the inside of [] is the theory of "Proto-consciousness".

* It is the root body [thing like a component] of the universe;

* It exists from prior of the Big Bang [from the Big Bang];

* Intention elements ["Quantum consciousness"] are connected by instantaneous long-range communication like quantum entanglement. [connected by quantum entanglement]

However, the concept of "Proto-consciousness" seems similar to the will of the universe creator or to the "collective unconscious" proposed by psychologist Jung.

On the other hand, it has the materialistic aspect that mind is born from a computer composed of microtubules which are substances.

Since it cannot be explained only by materialism, it seems to have taken in idealism.

There are some negative points like this, but when approaching the truth of the universe from the standpoint of a medical scientist, I can say that it has come closer to the "Proto-particle" (philosophical tetrahedron).

2-10. Elementary particles that transmit force

So far, we have excluded elementary particles that transmit forces such as bosons, but let's consider them here.

As shown on the right side of the vertical black line in the standard model of elementary particles shown in the beginning of Chapter 1, there are 4 kinds of elementary particles that transmit force.

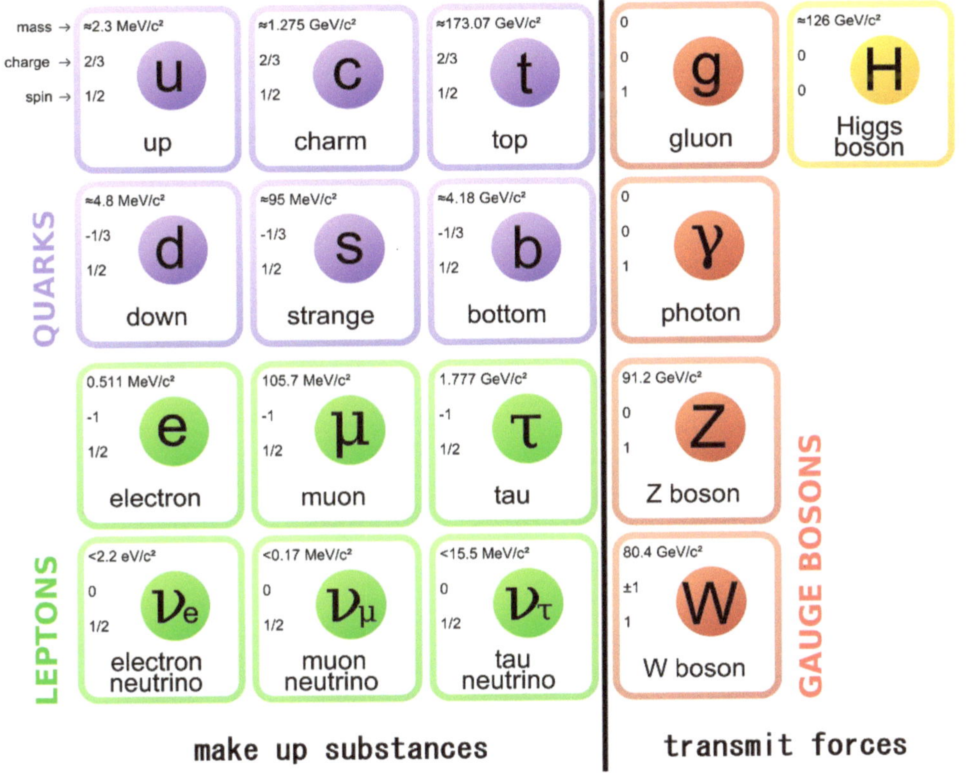

(Quoted from Wikipedia "Standard Model." I inserted the black vertical line there.)

As long as the root body of the universe is the "Proto-particle", these should also arise from the "Proto-particle".

Since faces of the philosophical tetrahedron, which is the "Proto-particle", becomes elementary particles that compose matter, elementary particles that transmit force may correspond to arms of the philosophical tetrahedron or combinations thereof.

In simple terms, it can be hypothesized that the action of only arms of the philosophical tetrahedron are observed as 4 kinds of elementary particles that transmit forces.

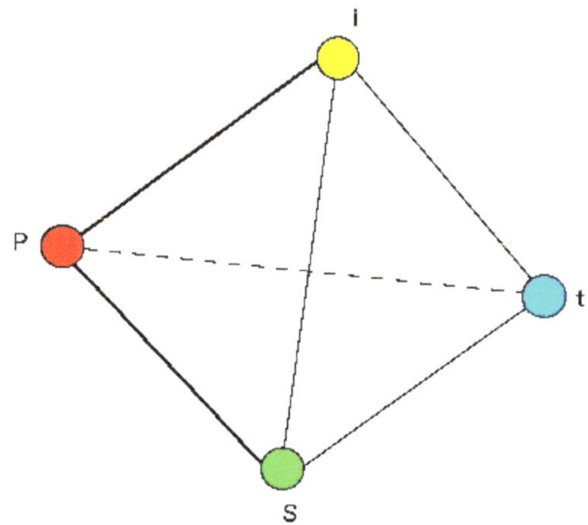

There are many combinations of arms, but it is said that there are 8 types of gluons, which are elementary particles that transmit force.

I'll just say that there are many possible combinations here.

3. Dark matter is the "Proto-particle"?

3-1. State where phenomenon does not occur "Kuu"

According to the 6-dimensional principle of Dr. Yamamoto, if the 4 fundamental elements of this world and the universe - space, time, power and intention - melt into one, it becomes a state called "Kuu".

Dr. Yamamoto described that they melt into one, but it means the state that there is no bias in any of the four elements.

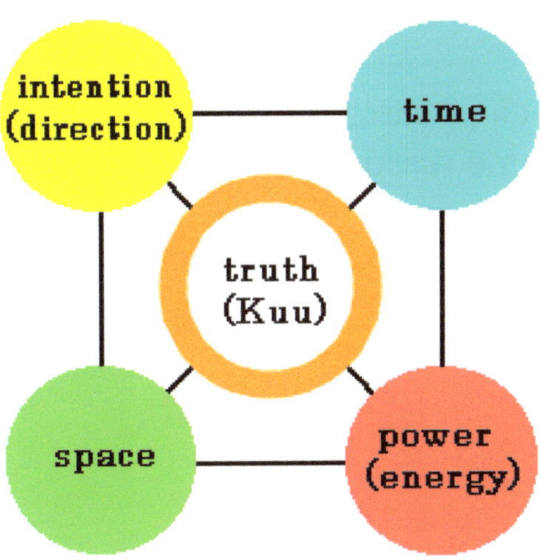

This word "Kuu" usually means empty, but here it's just a state that doesn't appear as a material phenomenon, and it is not completely nothingness.

In the 6-dimensional world, that is, in the universe, the state of "Kuu" and material phenomenon can occur reversibly.

Therefore, paranormal phenomena such as substances appearing or disappearing suddenly, which cannot be explained by conventional physics, can occur.

Dr. Yamamoto actually experienced many such phenomena and he called them collectively "Kuu"-ification (emptying) phenomenon.

3-2. The world of "Kuu", and Dark matter

Well, the philosophical tetrahedron (Proto-particle) in this "Kuu" state does not disappear, but is state that none of the four faces has become a phenomenon.

It is thought that a phenomenon occurs when the 4 elements are biased for some reason, and therefore the "Proto-particle" is materialized into an elementary particle.

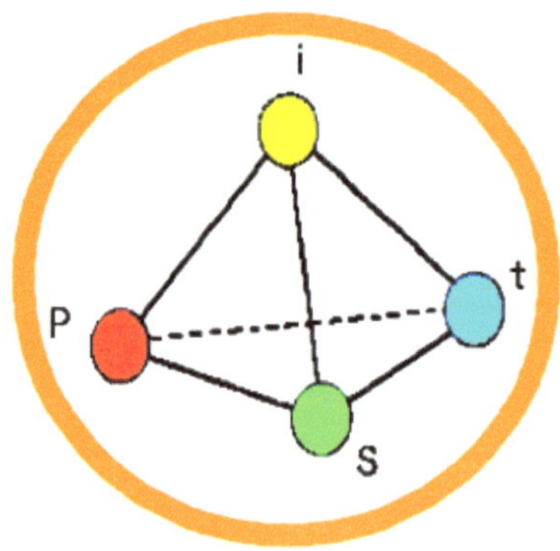

"Proto-particles" in the state of "Kuu" do not emit material energy as if there is nothing, so they do not react with others and the size cannot be observed.

However, only intention elements will keep the ability to communicate with others, otherwise the above-mentioned reversible "Kuu"-ification (emptying) phenomenon cannot occur. Also, if an intention element with memory is not working even in a state of "Kuu", once the object has been "Kuu"-ification (emptying), it will never be able to return to the former, and its reappearance will not occur.

And, if "Proto-particles" in the state of "Kuu" have the memory and communication capability of intention elements, then I think they will have gravity as described in chapter 2.

Inferring from that, "Proto-particles" in the state of "Kuu" may be mysterious substances called Dark matter, which exist in the universe, are affected by gravity, and are invisible.

3-3. Big Bang is explosion of "Proto-particles"?

Even if a large number of "Proto-particles" are in the state of "Kuu", which have no reactive and no size, gather together, they will still overlap infinitely without interfering with each other and remain in the state of "Kuu" as a whole.

If there is a bias for some reason, there is a possibility that it will become a phenomenon, that is, a materialization, in a chain reaction due to communication between intention elements.

I suspect that was Big Bang at the beginning of the universe.

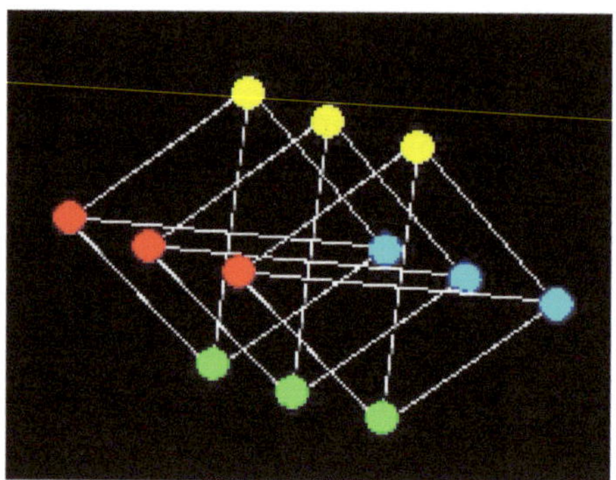

I think:

The space expands due to Big Bang, and "Proto-particles" are scattered to become the present universe. But not every "Proto-particles" were phenomenized and materialized, and many remained in the state of "Kuu". And they are recognized as Dark matter.

Therefore, in the space around us, and in the vacuum that seems to be nothing in the universe, "Proto-particles" are filled like air.

4. The universe will continue to expand?

The universe has continued to expand since Big Bang, but as mentioned earlier, if gravity is born from the memories of intentional elements that were once one, the universe may eventually begin to contract and gather in one lump.

Some scholars say that the universe is a living organism and is beating repeatedly expanding and contracting. Some say that even when the universe contracts to one point, neutrinos would still remain, and due to neutrinos, the universe turns into an expansion mode called inflation (see below figure). If neutrinos have a memory that once there was the universe, would that affect this?

Timeline of the universe
A representation of the evolution of the universe over 13.77 billion years

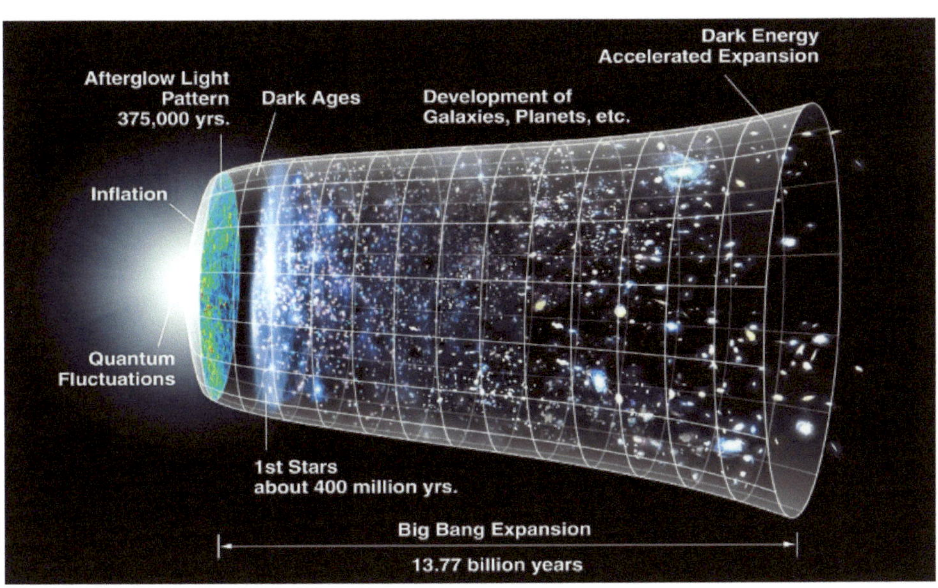

https://commons.wikimedia.org/wiki/File:CMB_Timeline300_no_WMAP.jpg

Afterword

In recent years, some scientists have used quantum random number generators to study how human consciousness acts on matter.

Dr. Dean Radin, a researcher in parapsychology from the Institute of Noetic Sciences, USA, observed that the unusual bias of several quantum random number generators occurred at the climax of burning a large doll in the Burning Man Festival which attracted 70,000 people.

Dr. Roger Nelson, a psychologist from Princeton University, USA, has been working on a research called "The Global Consciousness Project", whereby quantum random number generators are set at 50 locations in the world and kept under observation on a 24-hour basis. When the 9/11 terrorist attack in New York occurred in 2001, he observed anomalous bias of quantum random number generators in various parts of the world for several days.

These observations have accumulated data. However, the mechanism has not yet been elucidated.

As mentioned in my book (Mind Is Neutrino!? It Gives Evolution and Psychic Power), the element of mind may be neutrino. If so, the element of mind behaves like a quantum, and it seems to be the mechanism by which mind affects matter.

Also, recently, hypothetical theory that quantum entanglement produces gravity appears to have been issued. This is similar to the content of this book, "Gravity is occurred by the communication ability and their mutual call, of intention elements of the 'Proto-particles' (philosophical tetrahedra)".

I hope this book will be a clue to the elucidation of the universe.

January 2022 Hideo Asawa

Matsumoto City, Nagano Prefecture, Japan
Graduated from Tokyo University of Science

True Identity of Dark Matter!?
"Proto-Particle" Produce Elementary Particles

Published: March 2022

Author: Hideo Asawa

www.ingramcontent.com/pod-product-compliance
Ingram Content Group UK Ltd.
Pitfield, Milton Keynes, MK11 3LW, UK
UKHW060406300726
14090UKWH00006B/456

* 9 7 9 8 4 3 8 0 1 9 2 0 6 *